… **SCHOLASTIC**

PLAY & LEARN MATH

Addition & Subtraction

Learning Games and Activities to Help
Build Foundational Math Skills

by Mary Rosenberg

New York • Toronto • London • Auckland • Sydney
Mexico City • New Delhi • Hong Kong • Buenos Aires

Scholastic Inc. grants teachers permission to print and photocopy the reproducible pages from this book for classroom use. Purchase of this book entitles use of reproducibles by one teacher for one classroom only. No other part of this publication may be reproduced in whole or in part, or stored in a retrieval system, or transmitted in any form or by any means, electronic, mechanical, photocopying, recording, or otherwise, without written permission of the publisher. For information regarding permission, write to Scholastic Inc., 557 Broadway, New York, NY 10012.

Editor: Maria L. Chang
Cover design by Tannaz Fassihi
Cover art by Constanza Basaluzzo
Interior design by Grafica Inc.
Interior art by Mike Moran

Scholastic Inc., 557 Broadway, New York, NY 10012
ISBN: 978-1-338-31065-8
Copyright © 2019 by Mary Rosenberg
All rights reserved.
Printed in the U.S.A.
First printing, January 2019.
1 2 3 4 5 6 7 8 9 10   40   25 24 23 22 21 20 19

Contents

Introduction 5
Mathematics Standards Correlations 6

Adding to 5. 7
Word Problems to 5. 9
Picture Problems 11
Make the Number 14
Let's Subtract! 17
Subtraction Word Problems 22
Name That Addition Strategy 25
Switcheroo 28
Name That Subtraction Strategy 31
All or Nothing 34
Doubles Bingo 37
Fly Away Home 40
Add 'Em Up! 42
Fact Families 44
What's Missing? 46
Odds and Evens 50
Add and Subtract Odds and Evens 52
Make It Home 55
A Flip of the Cards 59
Duck Gallery 62

Introduction

As young children explore numbers and counting, they begin to get a sense of addition and subtraction. They realize that when they combine their crayons with a friend's, they end up with many more crayons. Or when they share their cookies with a sibling, they are left with fewer cookies. But before we can expect children to memorize addition and subtraction facts, we need to help them understand what these operations really mean.

Play & Learn Math: Addition & Subtraction helps solidify children's understanding of these key operations through a variety of fun games and activities. The activities require only a few simple classroom materials, such as number cubes (dice), colored counters, dominoes, and crayons. Many of the activities are written in two ways: one to guide teachers on how to implement the activity, and another to give children instructions using kid-friendly language. The purpose of this is simple—you, the teacher, introduce and model the activity to the whole class or to small groups, then children can use their own activity page, written at their level, as a step-by-step reminder of how to do the activity.

For kindergarteners, it is best to do these activities in a small-group setting. This way, you (or another adult, such as an instructional assistant or parent volunteer) can closely monitor each child and provide any needed correction, support, additional instruction, or practice. First and second graders can work independently or play the games with a partner after you have demonstrated how to do the activities to the whole class. As they engage in each activity, you can then walk around and provide additional support or coaching as needed.

While the activities are designed to meet specific grade-level standards (see Mathematics Standards Correlations, page 6), don't let this limit you to using only the activities in your grade level. Depending on your students' needs, feel free to move up or down a level to fill in any gaps in learning, to provide remediation on a specific skill, or to challenge more advanced learners.

Once children have played a game or practiced an activity several times, you can place it at a center. Children can then continue to practice previously learned skills independently or with a partner. Many of the activities can be increased in difficulty simply by adjusting the dice or numbers being used.

How can you tell which children need additional support and which ones need more challenging activities? One way to assess students' understanding is through the use of an "exit ticket." In each activity, you'll find a suggestion as to how you can adapt it as an exit ticket children need to complete before they go to recess, lunch, or home at the end of the day.

All set? Let's get playing!

Mathematics Standards Correlations*

Standard	Adding to 5	Word Problems to 5	Picture Problems	Make the Number	Let's Subtract!	Subtraction Word Problems	Name That Addition Strategy	Switcheroo	Name That Subtraction Strategy	All or Nothing	Doubles Bingo	Fly Away Home	Add 'Em Up!	Fact Families	What's Missing?	Odds and Evens	Add and Subtract Odds and Evens	Make It Home	A Flip of the Cards	Duck Gallery
KINDERGARTEN																				
OA.A.1: Represent addition and subtraction with objects, fingers, mental images, drawings, sounds (e.g., claps), acting out situations, verbal explanations, expressions, or equations.	✓	✓	✓	✓	✓	✓														
OA.A.2: Solve addition and subtraction word problems, and add and subtract within 10, e.g., by using objects or drawings to represent the problem.	✓	✓	✓	✓	✓	✓														
OA.A.3: Decompose numbers less than or equal to 10 into pairs in more than one way (e.g., by using objects or drawings), and record each decomposition by a drawing or equation (e.g., 5 = 2 + 3 and 5 = 4 + 1).	✓			✓																
OA.A.4: For any number from 1 to 9, find the number that makes 10 when added to the given number (e.g., by using objects or drawings), and record the answer with a drawing or equation.				✓																
OA.A.5: Fluently add and subtract within 5.	✓	✓	✓	✓	✓															
GRADE 1																				
OA.A.1: Use addition and subtraction within 20 to solve word problems involving situations of adding to, taking from, putting together, taking apart, and comparing, with unknowns in all positions e.g., by using objects, drawing, and equations with a symbol for the unknown number to represent the problem.		✓	✓			✓						✓								
OA.A.2: Solve word problems that call for addition of three whole numbers whose sum is less than or equal to 20, e.g., by using objects, drawings, and equations with a symbol for the unknown to represent the problem.													✓							
OA.B.3: Apply properties of operations as strategies to add and subtract. Examples: If 8 + 3 = 11 is known then 3 + 8 = 11 is also known. (Commutative property of addition.) To add 2 + 6 + 4, the second two numbers can be added to make a ten so 2 + 6 + 4 = 2 + 10 = 12. (Associative property of addition.)								✓					✓	✓						
OA.B.4: Understand subtraction as an unknown-addend problem. For example, subtract 10 − 8 by finding the number that makes 10 when added to 8.					✓										✓					
OA.C.5: Relate counting to addition and subtraction (e.g., by counting on 2 to add 2).				✓	✓	✓	✓	✓	✓			✓								
OA.C.6: Add and subtract within 20, demonstrating fluency for addition and subtraction within 10. Use strategies such as counting on; making ten; decomposing a number leading to a ten; using the relationship between addition and subtraction; and creating equivalent but easier or known sums (e.g., adding 6 + 7 by creating the known equivalent 6 + 6 + 1 = 13).			✓	✓	✓	✓	✓	✓	✓	✓	✓	✓	✓	✓	✓		✓	✓		✓
OA.D.8: Determine the unknown whole number in an addition or subtraction equation relating three whole numbers. For example, determine the unknown number that makes the equation true in each of the equations 8 + ? = 11, 5 = ___ − 3, 6 + 6 = ___.				✓	✓	✓								✓	✓			✓		✓
GRADE 2																				
OA.A.1: Use addition and subtraction within 100 to solve one- and two-step word problems involving situations of adding to, taking from, putting together, taking apart, and comparing, with unknowns in all positions, e.g., by using drawings and equations with a symbol for the unknown number to represent the problem.																	✓	✓		
OA.B.2: Fluently add and subtract within 20 using mental strategies. By end of Grade 2, know from memory all sums of two one-digit numbers.																				✓
OA.C.3: Determine whether a group of objects (up to 20) has an odd or even number of members, e.g., by pairing objects or counting them by 2s; write an equation to express an even number as a sum of two equal addends.																✓	✓			
OA.C.4: Use addition to find the total number of objects arranged in rectangular arrays with up to 5 rows and up to 5 columns; write an equation to express the total as a sum of equal addends.																			✓	

* © 2010 National Governors Association Center for Best Practices and Council of Chief State School Officers. All rights reserved.

Teacher Page

Adding to 5

This small-group activity is a great way to ease children into simple addition. They use counters of two different colors in various combinations to make 5. Then they color in a five frame and write the matching addition sentence.

HERE'S HOW

Divide the class into small groups of two to four children. Distribute copies of the "Adding to 5" activity page to children. Display a copy on the board.

Model how to do the activity. Make 5 using counters of two different colors. Place one counter in each space of the five frame until all spaces are filled. Keep the same colors together; for example, 3 red counters and 2 blue counters. (Don't split the colors to make patterns, such as red, blue, red, blue, red.) Next, color the squares on the activity sheet to match the counters. Finally, record the addition sentence on the lines below.

Have children in each group take turns making 5 and sharing it with the rest of their group.

MATERIALS

- Adding to 5 activity page (page 8)
- 8 counters in two colors (for each child)
- 2 crayons that match the colors of the counters (for each child)
- pencils
- classroom projection system

EXIT TICKET

Before leaving at the end of the day, give each child a number from 1 to 5. Have the child use fingers on both hands to show the target number. For example, say the target number is 4. The child might show 3 fingers on one hand and 1 finger on the other hand.

Activity Page

Name: _____ Date: _____

Adding to 5

Show different ways of making 5. Color in the boxes below. Then write the addition sentences.

1

_____ + _____ = _____

2

_____ + _____ = _____

3

_____ + _____ = _____

4

_____ + _____ = _____

Teacher Page

Word Problems to 5

This activity introduces children to addition word problems. Work with a small group of two or three children at a time, reading aloud the word problems to them. Then invite children to share their solutions with the group.

HERE'S HOW

Distribute copies of "Word Problems to 5" activity page to children.

Read aloud each word problem. On their activity page, have children circle the numbers used in each word problem. Then have them color in the squares to represent each type of animal in the word problem. (Use a different color for each type of animal.) Finally, ask children to write the addition sentence on the lines. Have children take turns sharing their answers with the group.

EXTENDED PRACTICE

Encourage children to make up their own word problems. Then invite each child to share his or her word problem with the group. Have children use counters to show and solve the problems.

MATERIALS

- Word Problems to 5 activity page (page 10)
- 2 different-color crayons (for each child)
- pencils

EXIT TICKET

Provide each small group of children with a dry-erase board and pen. Working as a group, have children develop a word problem to share with the rest of the class.

Activity Page

Name: _____ Date: _____

Word Problems to 5

Read or listen to each word problem. Color in the boxes to show the numbers. Then write the math sentence.

1. Sally has 2 dogs and 2 cats. How many pets does Sally have in all?

_____ + _____ = _____

2. Ben has 1 bird and 4 fishes. How many pets does Ben have in all?

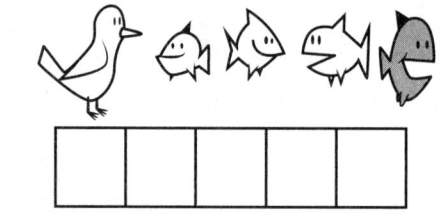

_____ + _____ = _____

3. Carson saw 2 frogs and 1 toad. How many animals did Carson see in all?

_____ + _____ = _____

4. Jane counted 1 zebra and 3 monkeys. How many zoo animals did Jane count?

_____ + _____ = _____

5. Mavis saw 3 ladybugs and 2 snails. How many animals did Mavis see in all?

_____ + _____ = _____

6. Tom counted 1 hen and 2 ducks. How many animals did Tom count?

_____ + _____ = _____

10

Teacher Page

Picture Problems

In this activity, children use picture cards to make up a short story problem and solve it. Not only does this activity give children practice in addition, it also encourages creativity and speaking to a group. Do this activity as a whole class or in small groups.

HERE'S HOW

Photocopy the "Picture Problems" picture cards and cut them apart. Make double-sided copies of the recording sheet (each page has the recording sheet on the front and back). Give each child a recording sheet and two different-color crayons.

Show children a picture card. Model how to tell a brief story about each picture.

For example:

Marcy picked 3 roses and 2 tulips. What does 3 and 2 equal?

Invite a child to select a picture card and tell a brief story about the picture. (Provide help and coaching as needed.) On their recording sheet, have children color in the squares to represent the numbers used in the word problem. Make sure they use a different color for each number. Then call on another child to tell the matching addition sentence. Repeat with the other picture cards.

EXTENDED PRACTICE

Provide children with old magazines, scissors, paper, and glue. Have them cut out pictures from the magazines and glue them to a sheet of paper. Then ask children to write a word problem to go with their picture. Invite children to share their word problems with the class.

MATERIALS

- Picture Problems picture cards (page 12)
- Picture Problems recording sheet (page 13)
- 2 different-color crayons (for each child)
- old magazines
- scissors
- glue
- paper

EXIT TICKET

Give each child a blank sheet of paper. Ask children to draw a picture and create their own word problem to share with the class.

Picture Cards

Picture Problems

Recording Sheet

Name: _____ Date: _____

Picture Problems

Pick a picture card. Tell a short story problem about it. Color in the squares to match the numbers. Then fill in the blanks to complete the sentence.

1

_____ and _____ equal _____ .

2

_____ and _____ equal _____ .

3

_____ and _____ equal _____ .

4

_____ and _____ equal _____ .

Teacher Page

Make the Number

With this activity, children learn there are several different ways to make a number. Provide children with linking cubes in two different colors to make a target number. Then have them write the matching addition sentence. Repeat several times, using different combinations of colored linking cubes and addition sentences to make the same target number.

HERE'S HOW

Write a target number (up to 10) in the top square on the "Make the Number" recording sheet then photocopy for children. (If you want to differentiate, photocopy the sheet beforehand then write a different number for each child.) Provide each child with several linking cubes in two different colors and two crayons that match the colors of the cubes.

Display a copy of the recording sheet on the board and model how to do the activity. Make the target number by combining two colors of the linking cubes. Then color in the boxes to match the colored cubes. Below the boxes, write an addition sentence to match the cubes.

For example:

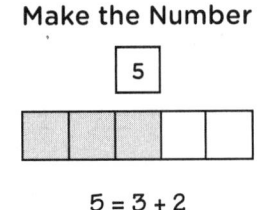

Make the Number

5

5 = 3 + 2

Have children write several different addition sentences for the target number. Provide help and coaching as needed. Then invite children to share their favorite addition sentence with the group.

EXTENDED PRACTICE

Making 10 is an important skill for children to know. Write "10" in the square at the top of the "Make the Number" recording sheet then photocopy for children. Provide each child with two different-color crayons and a number cube (die). To do the activity, have children roll the number cube and color in the matching number of squares on the recording sheet. They then use the second crayon to color in the remaining squares. Have them write the matching addition sentence under the squares. Repeat to complete the sheet.

MATERIALS

- Make the Number recording sheet (page 16)
- linking cubes (two colors for each child)
- 2 crayons that match the colors of the linking cubes (for each child)
- pencils
- classroom projection system
- number cube (die)

EXIT TICKET

Give the class a target number. Have each child tell you an addition sentence that has the target number as the sum.

Independent Practice

Make the Number

How many ways can you make the target number?

What to Do

1. Look at the target number. It's at the top of the sheet.

2. Make the target number. Use two colors of the linking cubes.

   For example, say the target number is 5. You might put 3 red cubes and 2 blue cubes together.

3. Color in the boxes to match the colored cubes.

4. Write an addition sentence that matches the boxes.

   For example, 5 = 3 + 2.

5. How else can you make the target number? Repeat Steps 2 to 4.

You'll Need

★ Make the Number recording sheet

★ linking cubes in 2 colors

★ 2 crayons that match the colors of the linking cubes

★ pencil

Recording Sheet

Name: _____ Date: _____

Make the Number

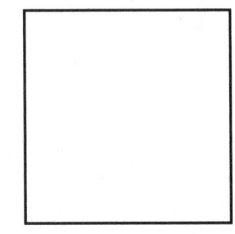

☐

1 ☐☐☐☐☐☐☐☐☐☐

_____ = _____ + _____

2 ☐☐☐☐☐☐☐☐☐☐

_____ = _____ + _____

3 ☐☐☐☐☐☐☐☐☐☐

_____ = _____ + _____

4 ☐☐☐☐☐☐☐☐☐☐

_____ = _____ + _____

16

Teacher Page

Let's Subtract!

Give children practice in subtraction with this fun and simple activity. Children pick number cards to make a subtraction problem. Then they color in circles and cross out the appropriate number to help solve the problem. Finally, they write the subtraction sentence. A great small-group activity for your math center!

HERE'S HOW

For each group of children, make two copies of the "Let's Subtract!" number cards and cut them apart.

Divide the class into small groups. Provide each group with two sets of the number cards and a card mat. Give each child a copy of the recording sheet and a crayon and pencil.

Model how to do the activity. Sort the number cards by their background color. Shuffle each set of cards. Place the cards facedown on the matching labeled box. (Cards with a white background go on the labeled white box, while cards with a black background go on the black box.) Next, turn over the top card in each stack and place the card in the space below it. Read aloud the subtraction problem made.

On the recording sheet, color in the number of circles to match the first card, then cross out the number of colored circles to match the second card. Then write the subtraction sentence on the blank lines.

For example:

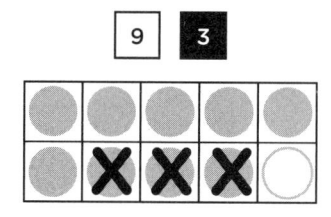

9 − 3 = 6

Have children take turns doing the steps until everyone in the group has completed their recording sheet.

MATERIALS

- Let's Subtract! number cards (page 19)
- Let's Subtract! card mat (page 20)
- Let's Subtract! recording sheet (page 21)
- crayons and pencils

EXIT TICKET

Using the number cards from the activity, show each child two cards. Ask the child to tell you the subtraction problem and its answer.

Independent Practice

Let's Subtract!

Pick two number cards. Subtract the numbers.

What to Do

1 Sort the number cards by color. Shuffle each set of cards.

2 Place the cards facedown on the card mat. (White cards go on the labeled white box. Black cards go on the black box.)

3 Turn over the top card in each stack. Place each card in the space below it.

4 Read aloud the subtraction sentence you made.

5 On your recording sheet, color in the number of circles to match the first card.

6 Then cross out the number of colored circles to match the second card.

7 Write the subtraction sentence under the circles.

8 Repeat Steps 3 to 7 two more times.

You'll Need

- ★ Let's Subtract! number cards
- ★ Let's Subtract! card mat
- ★ Let's Subtract! recording sheet (for each child)
- ★ crayons and pencils

Number Cards

Let's Subtract!

10	9	8
7	6	5
5	4	3
2	1	0

Card Mat

Let's Subtract!

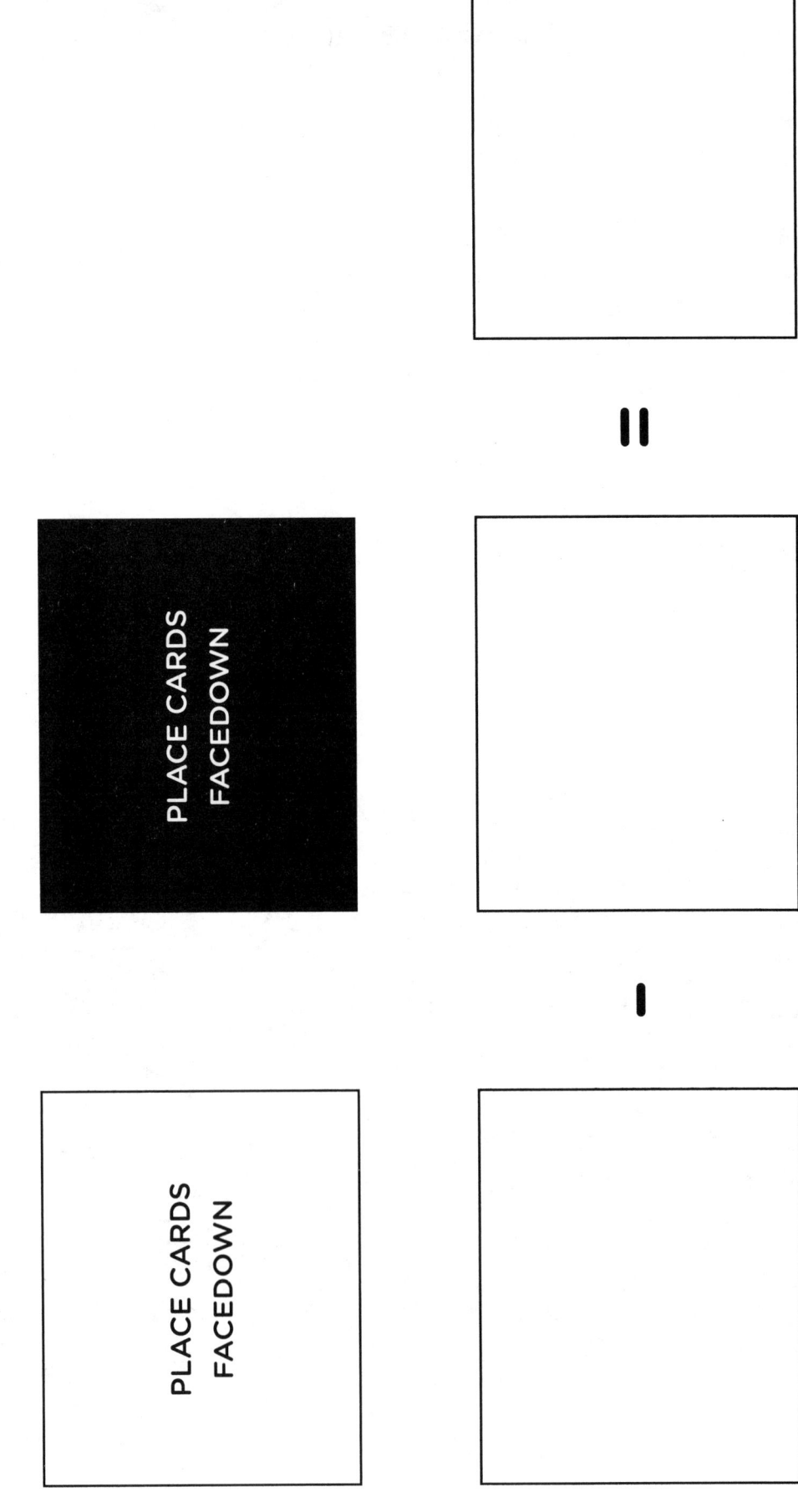

[PLACE CARDS FACEDOWN]

[   ]
−
[   ]
=
[   ]

[PLACE CARDS FACEDOWN]

Recording Sheet

Name: _____ Date: _____

Let's Subtract!

Pick two number cards. Color the number of circles to match the first card.
Cross off the number of colored circles to match the second card.
Write the subtraction sentence.

1

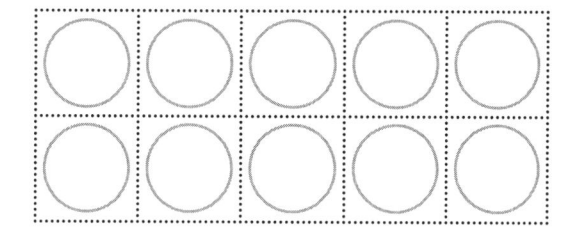

_____ - _____ = _____

2

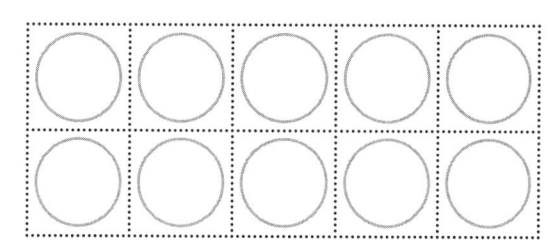

_____ - _____ = _____

3

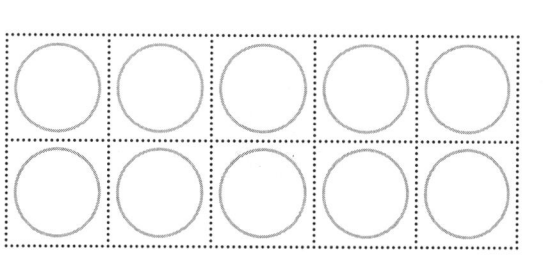

_____ - _____ = _____

Teacher Page

Subtraction Word Problems

Help children tackle subtraction word problems with this whole-class (or small-group) activity. Model how to draw circles and cross them out to help children visualize each problem and solve it.

HERE'S HOW

Photocopy the "Subtraction Word Problems" cards and cut them apart. Make double-sided copies of the recording sheet (each page has the recording sheet on the front and back). Give each child a copy of the recording sheet and a pencil.

Pick the first word problem and read it aloud to children. Model how to draw a quick picture to show the word problem. Draw 8 circles for the birds, then cross out 3 circles to show the number subtracted. Then write the subtraction sentence to go with the picture: 8 − 3 = 5. Have children do the same on their recording sheet.

Working together, read the next word problem to children. Ask them:

  a. *How many kittens were there to begin with?* (10)

  b. *How many kittens started playing?* (4)

  c. *How many kittens should we draw?* (10)

  d. *How many kittens should we cross off?* (4)

  e. *What is the subtraction sentence?* (10 − 4 = 6)

Continue with the other subtraction word problems. Monitor children as they work to make sure they are drawing the appropriate pictures and writing corresponding subtraction sentences.

MATERIALS

- Subtraction Word Problems cards (page 23)
- Subtraction Word Problems recording sheet (page 24)
- pencils

EXIT TICKET

On a blank sheet of paper, have each child draw a subtraction problem using simple items, such as apples, dog bones, cupcakes, or smiley faces. (Children should draw a number of items and cross some of them off.) Using their drawings, have children orally share their subtraction problems.

Problem Cards

Subtraction Word Problems

1.
There were 8 birds in a tree.
3 birds flew away.
How many birds are left in the tree?

2.
There were 10 kittens napping.
4 kittens woke up and played.
How many kittens are still napping?

3.
There were 9 kids eating lunch.
7 kids finished eating.
How many kids are still eating lunch?

4.
There were 8 children picking out books.
1 child started reading.
How many children are still picking out books?

5.
There were 6 teddy bears at the tea party. 3 teddy bears left. How many teddy bears are still at the tea party?

6.
Mavis planted 7 seeds in a pot.
2 seeds started sprouting.
How many seeds have not sprouted?

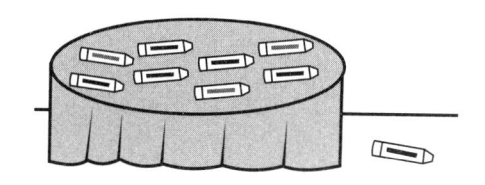

7.
John had 9 crayons on the table.
1 crayon rolled onto the floor.
How many crayons are still on the table?

8.
There were 4 kids running.
1 kid stopped to tie his shoe.
How many kids are still running?

23

Recording Sheet

Name: _____ Date: _____

Subtraction Word Problems

_____ - _____ = _____

_____ - _____ = _____

_____ - _____ = _____

_____ - _____ = _____

Teacher Page

Name That Addition Strategy

Help children identify, review, and practice various addition strategies with this game. Children can play this game in pairs, in small groups, or on their own.

HERE'S HOW

Review the different addition strategies with children:

- **Count on 1 (or 2 or 3):** Children start with the greater number and then count on 1 (or 2, or 3). For example, if they roll a 5 and a 2, they start with 5 and count up 2: "5 . . . 6, 7." So 5 + 2 = 7.

- **Double:** Explain to children that "double" means having two of the same. So if you have two 1s, that's 1 + 1 = 2. If you have two 2s, that's 2 + 2 = 4.

- **Double + 1:** This strategy will come easily once children know their doubles. Explain that if one number is one greater than another, for example 3 and 4, then they can use their double facts to say, "3 + 3 = 6, so 3 + 4 is just one more. 3 + 4 = 7."

After you have reviewed the strategies, divide the class into small groups or pairs. Provide each group with two number cubes, then give each child a copy of the "Name That Addition Strategy" recording sheet and a pencil.

Model how to play this game. Roll both number cubes. Think aloud about which strategy is most appropriate for solving the addition problem. Then write the addition sentence in the appropriate column on the recording sheet.

For example:

Count on 1	Count on 2	Count on 3	Double	Double + 1	Other
			3 + 3 = 6		

Have children take turns rolling the number cubes and filling in their recording sheet. The first child to complete a column wins the game.

MATERIALS

- Name That Addition Strategy recording sheet (page 27)
- 2 number cubes (dice)* for each pair or small group of children
- pencils

* **Variation:** For children who are ready for a challenge, provide 9- or 12-sided dice.

EXIT TICKET

Use a poster machine to make a large copy of the recording sheet. Laminate the sheet so it can be reused. Have each child roll two number cubes. Using a dry-erase or water-based marker, have the child record the resulting addition sentence in the appropriate column on the poster.

Name That Addition Strategy

What addition strategies do you know?

What to Do

1. Roll both number cubes.

2. Add the numbers. Which strategy did you use to add them?

3. On your recording sheet, look for the strategy you used. Write the addition sentence under that strategy.

4. The next player takes a turn.

5. Continue taking turns. The first player to complete a column wins.

Players: 2 to 4

You'll Need

★ Name That Addition Strategy recording sheet (for each child)

★ 2 number cubes

★ pencil (for each child)

Recording Sheet

Name: _____ Date: _____

Name That Addition Strategy

Count on 1	Count on 2	Count on 3	Double	Double + 1	Other

Teacher Page

Switcheroo

Dominoes present a great way to introduce children to the commutative property of addition. The simple act of rotating a domino will help children understand that even when the numbers are switched around, their sum will remain the same.

HERE'S HOW

Divide the class into pairs or into small groups of three or four children. Give each group a set of double-six dominoes. Distribute copies of the "Switcheroo" recording sheet to children.

Model how to do the activity. Turn the dominoes facedown and mix them around. Take a domino and draw the dots on the top box of the recording sheet. Next, rotate the domino and draw the dots on the bottom box. Then write the matching addition sentences on the lines below.

For example:

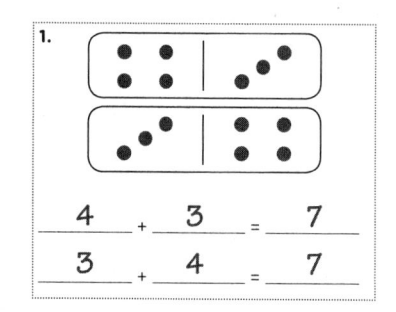

$$4 + 3 = 7$$
$$3 + 4 = 7$$

Have children take turns picking a domino and recording on their sheet. They should continue taking turns until they fill up their recording sheets.

MATERIALS

- Switcheroo recording sheet (page 30)
- set of double-six dominoes (for each pair or small group of children)
- pencils

EXIT STRATEGY

Show each child a domino. Have the child tell you the two addition sentences that can be made using the domino.

28

Independent Practice

Switcheroo

Pick a domino. Use it to write two addition sentences.

What to Do

1. Turn the dominoes facedown on the table. Mix them around.

2. Pick a domino. Turn it over.

3. Draw the dots in the top box of your recording sheet.

4. Turn the domino around. Draw the dots in the bottom box.

5. Write the addition sentences below.

6. Repeat Steps 2 to 5 a few more times.

You'll Need
- ★ Switcheroo recording sheet (for each child)
- ★ dominoes
- ★ pencil (for each child)

Recording Sheet

Name: _____ Date: _____

Switcheroo

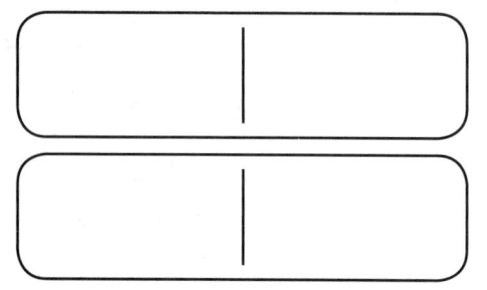

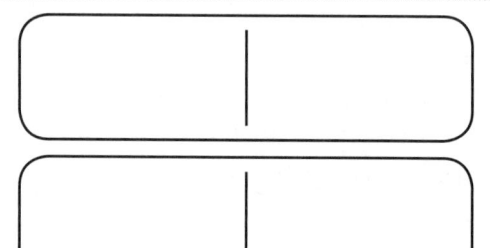

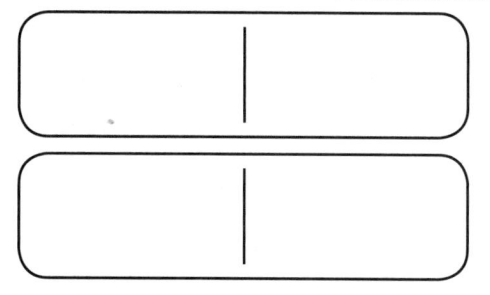

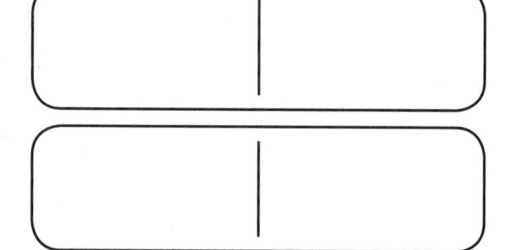

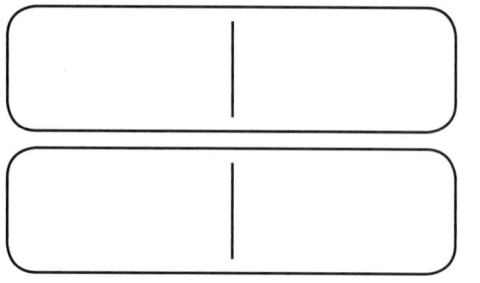

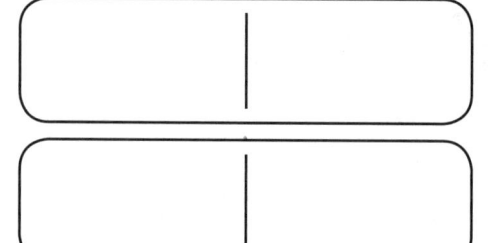

1.

____ + ____ = ____

____ + ____ = ____

2.

____ + ____ = ____

____ + ____ = ____

3.

____ + ____ = ____

____ + ____ = ____

4.

____ + ____ = ____

____ + ____ = ____

5.

____ + ____ = ____

____ + ____ = ____

6.

____ + ____ = ____

____ + ____ = ____

Teacher Page

Name That Subtraction Strategy

Similar to "Name That Addition Strategy" (page 25), this game helps children identify, review, and practice various subtraction strategies. Children can play this game in pairs, in small groups, or on their own.

HERE'S HOW

Review the different subtraction strategies with children, giving examples.

- **Subtract all:** This may be the easiest subtraction strategy for children to remember. Explain that when you subtract a number by itself—in other words, subtract all—you're taking all away, so the result is always 0. For example, if you take 2 away from 2, what's left is 0, so 2 – 2 = 0.

- **Count back 1 (or 2 or 3):** Children start with the greater number and then count back 1 (or 2, or 3). For example, if they roll a 4 and a 3, they start with 4 and count down 3: "4 . . . 3, 2, 1." So 4 – 3 = 1.

After you have reviewed the strategies, divide the class into small groups or pairs. Provide each group with two number cubes, then give each child a copy of the "Name That Subtraction Strategy" recording sheet and a pencil.

Model how to play the game. Roll both number cubes. Think aloud about which strategy is most appropriate for solving the subtraction problem. Then write the subtraction sentence in the appropriate column on the recording sheet. (Remind children when subtracting to always start with the greater number.)

Children take turns rolling the number cubes and filling in their recording sheet. The first child to complete a column wins the game.

For example:

Subtract All	Count Back 1	Count Back 2	Count Back 3	Other
		5 – 2 = 3		

MATERIALS

- Name That Subtraction Strategy recording sheet (page 33)
- 2 number cubes (dice) for each pair or small group of children
- pencils

EXIT TICKET

For each child, name a strategy (Subtract All, Count Back 1, Count Back 2, or Count Back 3), then ask the child to share a subtraction sentence that fits the strategy.

Game Directions

Name That Subtraction Strategy

What subtraction strategies do you know?

What to Do

1. Roll both number cubes.

2. Subtract the numbers. (Remember: Start with the greater number.) Which strategy did you use to subtract?

3. On your recording sheet, look for the strategy you used. Write the subtraction sentence under that strategy.

4. The next player takes a turn.

5. Continue taking turns. The first player to complete a column wins.

Players: 2 or 4

You'll Need

★ Name That Subtraction Strategy recording sheet (for each child)

★ 2 number cubes

★ pencil (for each child)

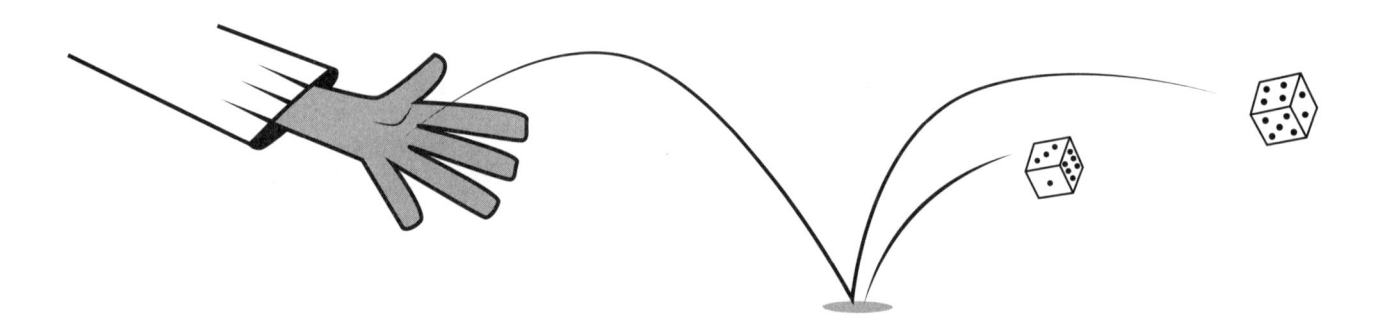

Recording Sheet

Name: _____ Date: _____

Name That Subtraction Strategy

Subtract All	Count Back 1	Count Back 2	Count Back 3	Other

All or Nothing

Children practice two important subtraction strategies with this quick and easy activity. Perfect for math centers or independent practice.

HERE'S HOW

Pair up children and give each pair a copy of the "All or Nothing" recording sheet, along with two paper clips and a pencil. Show children how to make a spinner using the paper clip and pencil. Hold the paper clip in place with a pencil over the spinner's center, as shown. Use a finger to flick the paper clip.

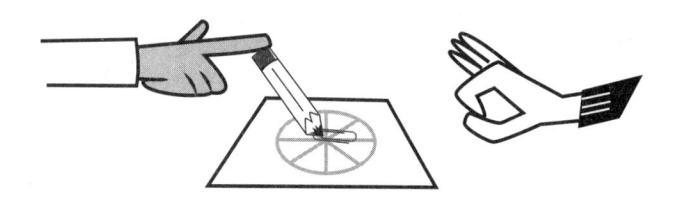

Model how to do the activity. Spin the two spinners on the recording sheet, then write the matching subtraction problem.

For example:

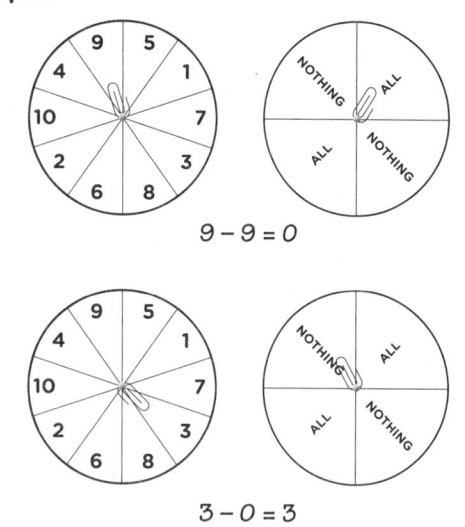

$9 - 9 = 0$

$3 - 0 = 3$

Have children take turns with the spinners until they complete their recording sheets.

MATERIALS

- All or Nothing recording sheet (page 36)
- paper clips
- pencils

EXIT TICKET

Tell each child a number and then say either "subtract all" or "subtract nothing." The child then tells you the subtraction problem and answer. For example, if you say, "4 and subtract all," the child should respond, "4 – 4 = 0."

Game Directions

All or Nothing

Spin a number. Then spin ALL or NOTHING. Write the matching subtraction sentence.

What to Do

1 Practice how to use the spinner. Hold the paper clip and pencil, as shown. Flick the paper clip with your finger.

Players: 2

You'll Need

★ All or Nothing recording sheet

★ pencils and paper clips (to make spinners)

2 On the recording sheet, spin the spinner on the left to get a number.

3 Spin the spinner on the right to get ALL or NOTHING.

4 Write the matching subtraction sentence.

For example:

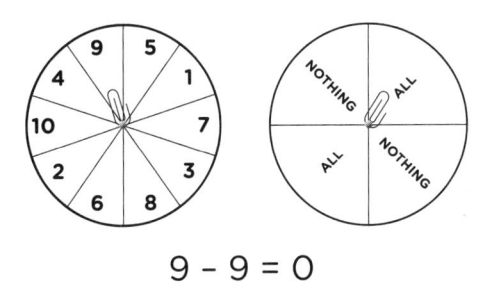

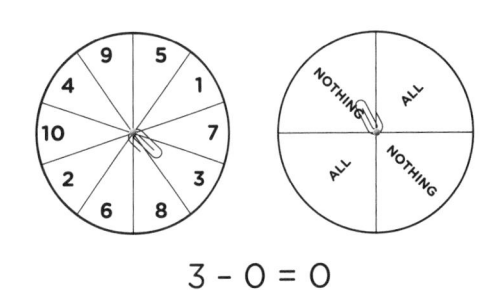

9 − 9 = 0   3 − 0 = 0

5 The next player takes a turn. Continue taking turns to fill up the recording sheet.

35

Recording Sheet

All or Nothing

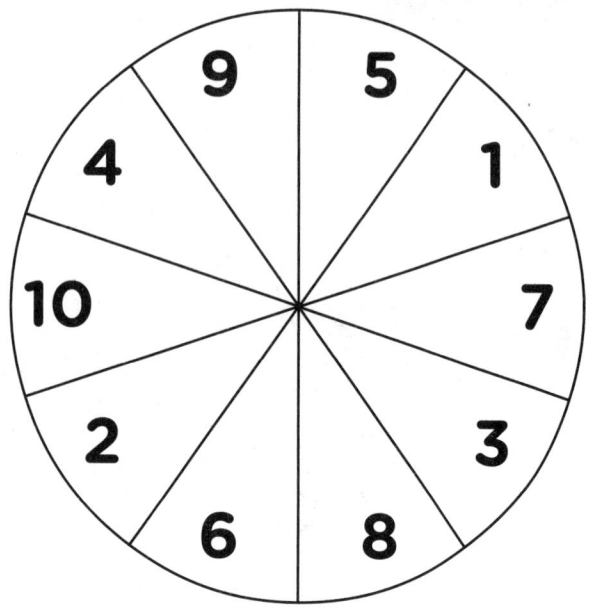

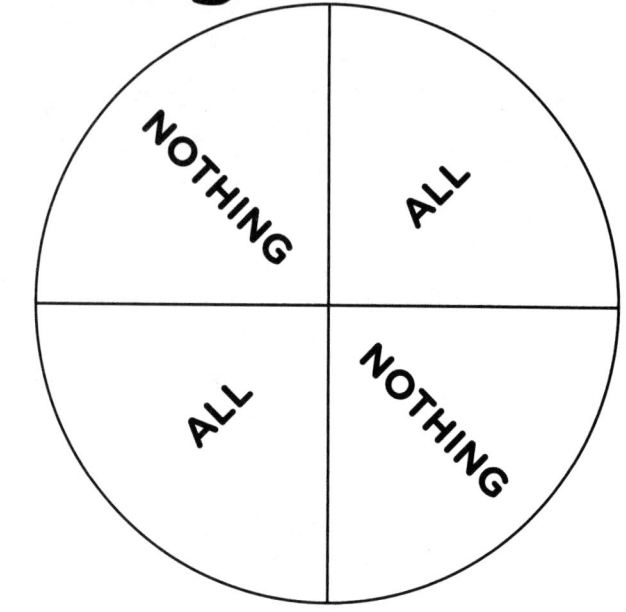

Player 1:	Player 2:
_____	_____
Spin #1: _____	Spin #1: _____
Spin #2: _____	Spin #2: _____
Spin #3: _____	Spin #3: _____
Spin #4: _____	Spin #4: _____
Spin #5: _____	Spin #5: _____
Spin #6: _____	Spin #6: _____
Spin #7: _____	Spin #7: _____
Spin #8: _____	Spin #8: _____
Spin #9: _____	Spin #9: _____
Spin #10: _____	Spin #10: _____

Teacher Page

Doubles Bingo

By playing this variation of Bingo, children build fluency with their doubles facts. Rolling a number cube, children double the number and find the sum on their Bingo cards. A fun, learning-filled game for pairs or small groups of children.

HERE'S HOW

Pair up children or divide the class into small groups. Provide each group with a number cube, then give each child a copy of the "Doubles Bingo" playing card and a pencil.

Model how to play the game. Roll the number cube. Using the number rolled, make a double addition problem. Write the addition sentence under a matching sum on the playing card. For example, say you roll a 3. Write 3 + 3 = 6 in one of the boxes labeled 6.

Playing Card

Doubles Bingo

10	8	4	2
6	2	10	12
12	6	8	4
4	8	2	6 (3 + 3 = 6)

Players take turns rolling the number cube. The first player to write four double addition sentences in a row—vertically, horizontally, or diagonally—wins the game.

MATERIALS

- Doubles Bingo playing card (page 39)
- number cubes (dice)
- pencils

EXIT TICKET

Tell each child a number from 0 to 10 and have the child tell you the doubles addition sentence. For example, if you say, "9," the child responds, "9 + 9 = 18."

To make the Exit Ticket more challenging, give each child a doubles sum, such as 16. The child then gives the doubles fact that makes 16 (8 + 8 = 16).

Game Directions

Doubles Bingo

Make double addition sentences to win Bingo!

What to Do

1 Players take turns rolling the number cube.

2 At your turn, use the number to make a double addition problem.

For example, if you roll a 3, the problem would be 3 + 3.

3 Look for the sum in your Doubles Bingo card.

4 Write the addition sentence under the sum.

5 The next player takes a turn.

6 Continue taking turns.

7 The first player to fill four boxes in a row—up and down, across, or diagonally—wins.

Players: 2 to 4

You'll Need

★ Doubles Bingo playing card (for each child)

★ pencil (for each child)

★ number cube

Playing Card

Doubles Bingo

10	8	4	2
6	2	10	12
12	6	8	4
4	8	2	6

Teacher Page

Fly Away Home

A hive of busy bees provides the setting for children to practice subtraction from 20. On the activity sheet, write a starting number from which to subtract. You may want to give the whole class the same number or differentiate by giving each child a different number. Allow children to work with a partner or on their own.

HERE'S HOW

Write a starting number (from 11 to 20) on the hive on the "Fly Away Home" activity page, then photocopy for children. (If you want to differentiate, photocopy the sheet beforehand, then write a different number for each child.) Provide each child with a crayon, paper clip, and pencil. Show children how to make a spinner using the paper clip and pencil. Hold the paper clip in place with a pencil over the spinner's center, as shown. Use a finger to flick the paper clip.

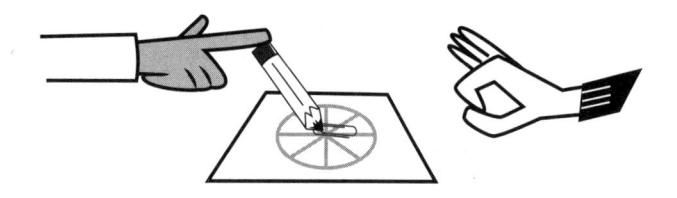

Display a copy of the activity page on the board and model how to do the activity. On the activity page, color the number of boxes that matches the starting number. Spin the spinner and subtract the number it lands on (cross out the matching number of boxes). Write the subtraction sentence underneath the boxes.

For example:

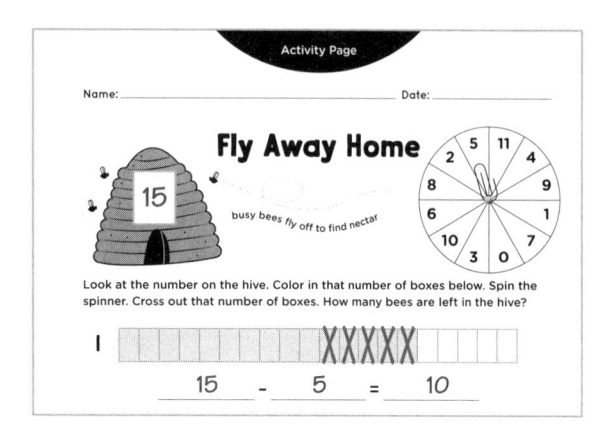

Have children repeat to complete the activity page.

MATERIALS

- Fly Away Home activity page (page 41)
- paper clips
- pencils
- crayons
- classroom projection system

EXIT TICKET

Give each child a starting number from 11 to 20 and have him or her spin the spinner on the recording sheet. Then have the child subtract the number shown on the spinner and tell the subtraction problem and answer. For example, say the starting number is 14 and the spinner shows 7. The child says, "14 – 7 = 7." (Note: You can use a different starting number for each child.)

ACTIVITY Page

Name: _____ Date: _____

Fly Away Home

busy bees fly off to find nectar

Spinner: 5, 11, 2, 4, 8, 9, 6, 1, 10, 7, 3, 0

Look at the number on the hive. Color in that number of boxes below. Spin the spinner. Cross out that number of boxes. How many bees are left in the hive?

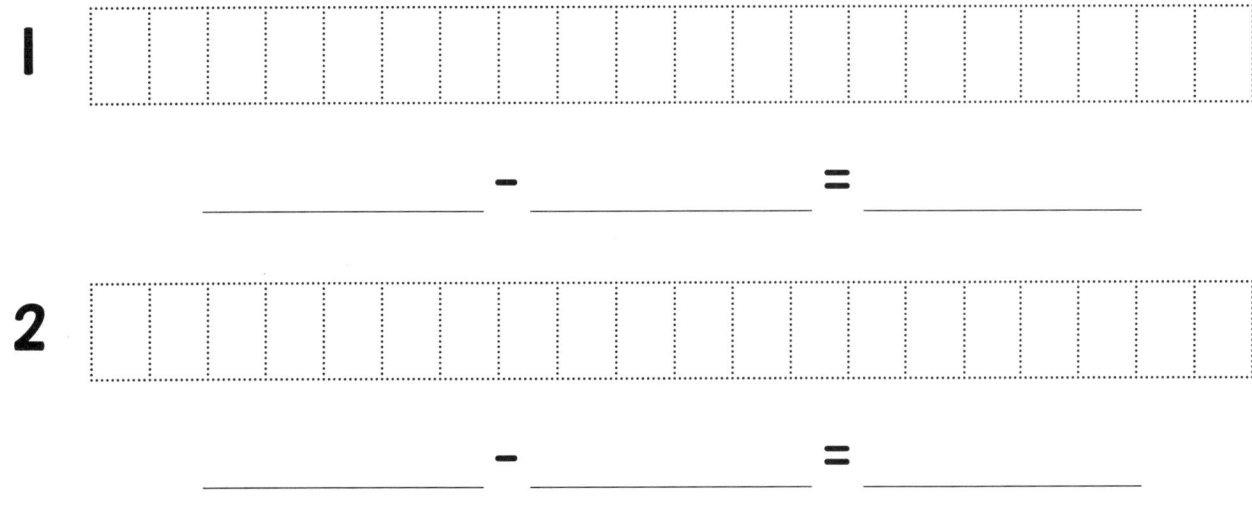

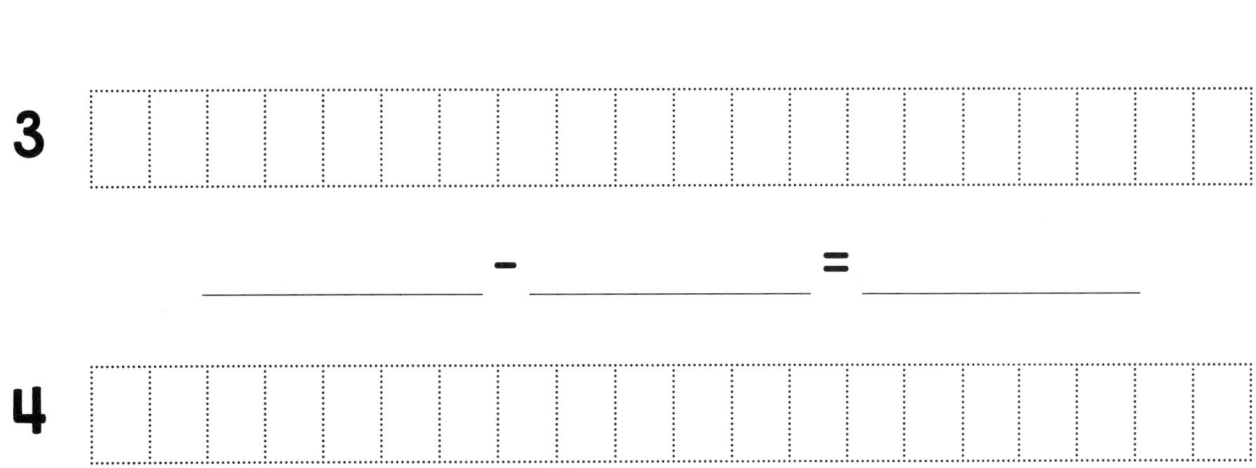

1

_____ - _____ = _____

2

_____ - _____ = _____

3

_____ - _____ = _____

4

_____ - _____ = _____

41

Teacher Page

Add 'Em Up!

Introduce children to the associative property of addition with this activity. Using number cubes to get three addends, children group the numbers in two different ways to find their sum. After modeling the activity, have children work in pairs or independently to complete their activity pages.

HERE'S HOW

Distribute copies of the "Add 'Em Up!" activity page to children. Give each child (or pair of children) three number cubes. (Alternatively, give each child one number cube to roll three times.)

Display the activity page on the board and model how to do the activity. Roll the three number cubes and write the numbers in the boxes on the activity page. Add the first two numbers together and write the sum in the box below them. Then write the third number and add both numbers together. Next, write the first number in the first box. Then add the second and third numbers together and write the sum in the box below them. Then add both numbers together to get the sum.

Have children work with a partner or on their own to complete their activity pages. As children work, walk around to monitor and provide help and coaching as needed.

MATERIALS

- Add 'Em Up! activity page (page 43)
- 3 number cubes (dice) for each child or pair of children
- pencils
- classroom projection system

EXIT TICKET

On index cards, write the numbers 0 to 9. Show each child three different cards. Have the child add two numbers together, then add the third number. For example, say you show the numbers 4, 3, and 7. The child says, "4 + 3 = 7. Then 7 + 7 = 14."

Name: _____   Date: _____

Add 'Em Up!

Roll three number cubes. Write the numbers in the top boxes. Then add the numbers in two different ways.

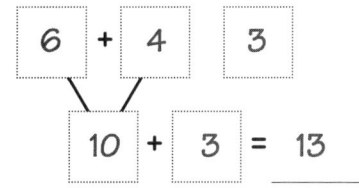

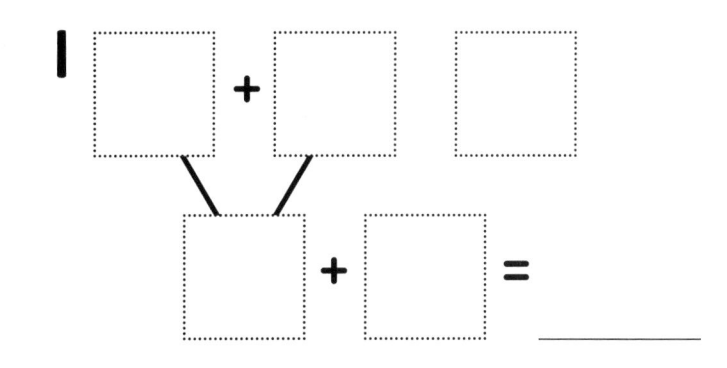

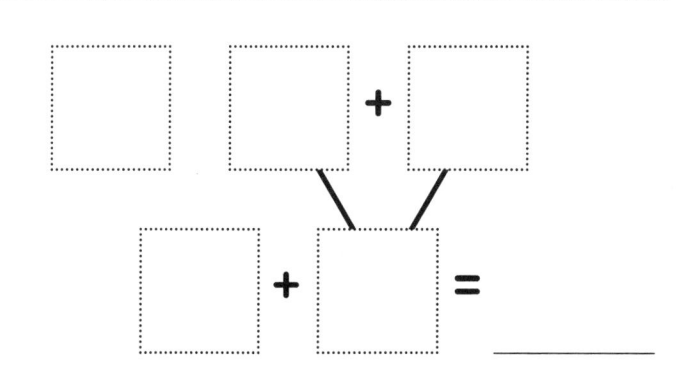

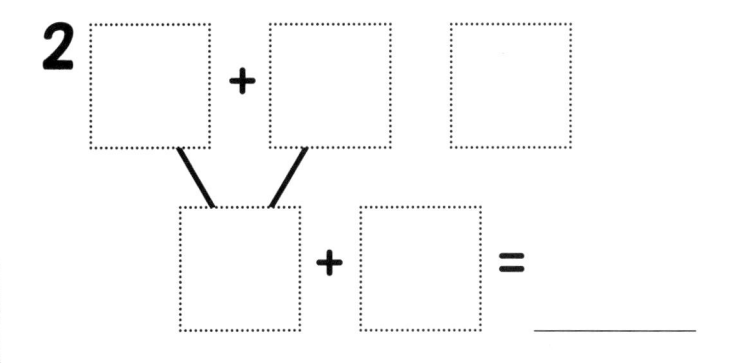

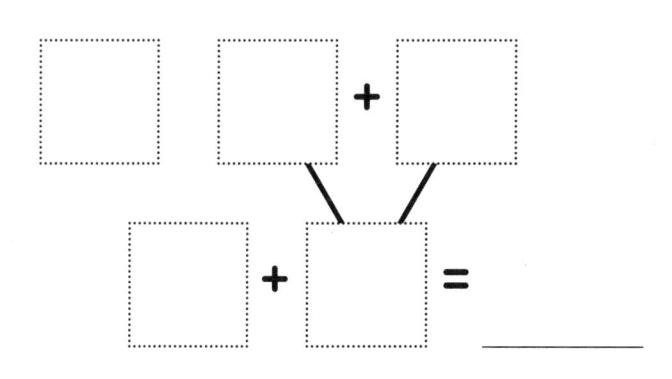

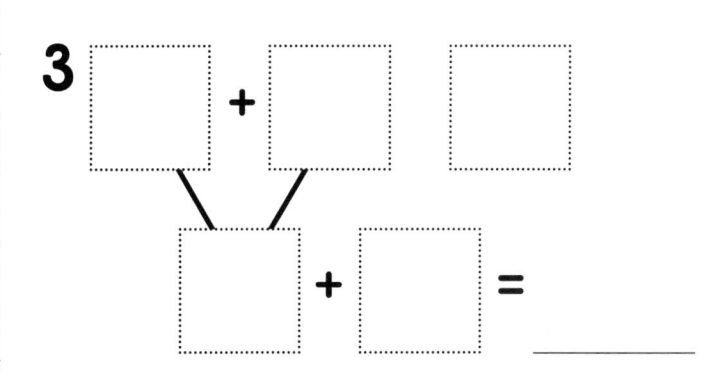

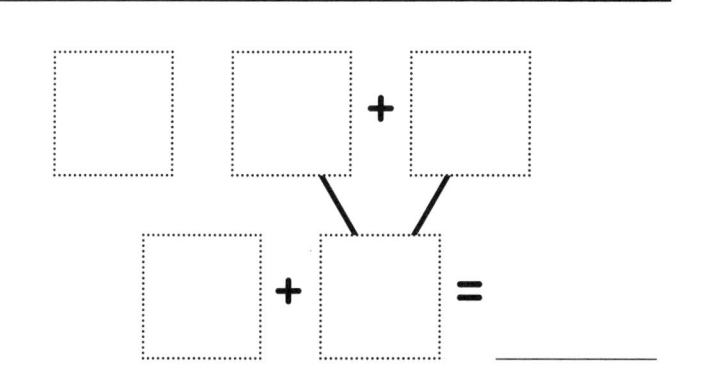

Example: 6 + 4   3   6   4 + 3
           10 + 3 = 13   6 + 7 = 13

Teacher Page

Fact Families

Fact families are a great way to master addition and subtraction facts. Children may be surprised to learn that they can create four math sentences out of just three numbers. You may want to have children do this activity with a partner first, then independently.

HERE'S HOW

Distribute copies of the "Fact Families" activity page to children. Provide each child (or pair of children) with two number cubes.

Display the activity page on the board and model how to do the activity. Roll the number cubes and record the numbers in the first two boxes. Add the two numbers and record the sum in the third box. Then write the fact families in the spaces below by writing two addition sentences and two subtraction sentences using the three numbers. Remind children that when writing the subtraction problems, they must start with the greatest number first.

Have children work independently or with a partner to write the different fact families. Provide help and coaching as needed.

MATERIALS

- Fact Families activity page (page 45)
- 2 number cubes (dice), for each child or pair of children
- pencils
- classroom projection system

EXIT TICKET

Give each child two numbers. Have the child say an addition sentence and a subtraction sentence that can be made using the numbers. For example, say you give a child the numbers 7 and 5. The child could then say, "7 + 5 = 12 and 12 − 5 = 7."

Name: _____   Date: _____

Fact Families

Roll two number cubes. Write the numbers in the first two boxes. Add them to get the third number. Then write the fact families below.

1. ☐ ☐ ☐

    ___ + ___ = ___
    ___ + ___ = ___
    ___ − ___ = ___
    ___ − ___ = ___

2. ☐ ☐ ☐

    ___ + ___ = ___
    ___ + ___ = ___
    ___ − ___ = ___
    ___ − ___ = ___

3. ☐ ☐ ☐

    ___ + ___ = ___
    ___ + ___ = ___
    ___ − ___ = ___
    ___ − ___ = ___

4. ☐ ☐ ☐

    ___ + ___ = ___
    ___ + ___ = ___
    ___ − ___ = ___
    ___ − ___ = ___

5. ☐ ☐ ☐

    ___ + ___ = ___
    ___ + ___ = ___
    ___ − ___ = ___
    ___ − ___ = ___

6. ☐ ☐ ☐

    ___ + ___ = ___
    ___ + ___ = ___
    ___ − ___ = ___
    ___ − ___ = ___

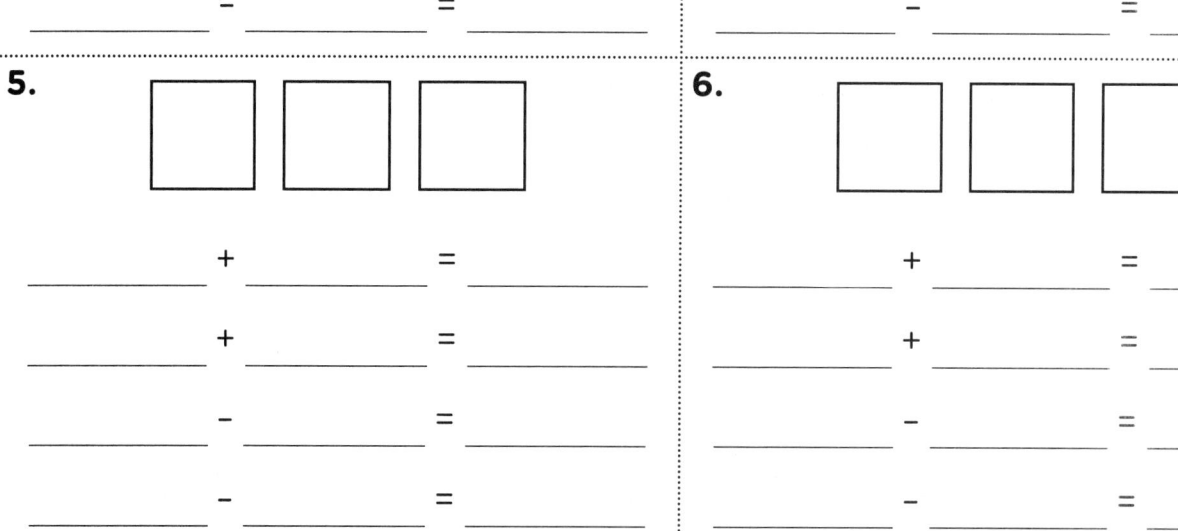

Teacher Page

What's Missing?

This two-player card game requires children to identify the missing number in a math problem—an important algebraic thinking skill. Working with both addition and subtraction problems, children must use various strategies to find the answer.

HERE'S HOW

For each pair of children, make two sets of the "What's Missing?" playing cards and cut them apart.

Pair up children. Provide each pair with two sets of the "What's Missing?" playing cards and a copy of the "What's Missing?" recording sheet.

Model for children how to play the game. Shuffle the cards and stack them on the table. Turn over the top card and read the math problem with the blank. Identify what number is needed to complete the math sentence. Write the math sentence next to the missing number on the recording sheet.

For example:

The missing number is 1. Write the math problem next to the 1 on the recording sheet.

3	
2	
1	1 + 9 = 10

Card: ___1___ + 9 = 10

Have children take turns turning over a card and writing the math problem on their side of the recording sheet. The goal is to have one math problem for each number. The first child to fill in all of the numbers wins the round.

MATERIALS

- What's Missing? playing cards (page 48)
- What's Missing? recording sheet (page 49)
- pencils

EXIT TICKET

Give each child an addition or subtraction problem with a missing number. Have the child identify the missing number and then restate the problem. For example, if you say, "3 + ___ = 10," then the child responds, "3 + 7 = 10." (Note: You can use the playing cards from the activity.)

Game Directions

What's Missing?

A number is missing in each math sentence. Can you name that number?

What to Do

1 Shuffle the cards. Stack them facedown on the table.

2 Turn over the top card. Read the math problem with the blank.

3 Think: What number is missing? Find that number on the recording sheet.

4 Write the complete math sentence next to the number that was missing.

5 The next player takes a turn. Repeat Steps 2 to 4.

6 Continue taking turns. The first player to complete his or her side of the sheet wins.

Players: 2

You'll Need

★ What's Missing? playing cards

★ What's Missing? recording sheet

★ pencil (for each child)

Playing Cards

What's Missing?

___ + 9 = 10	2 + 8 = ___	3 + ___ = 10	___ + 6 = 10
5 + ___ = 10	1 + 1 = ___	7 + ___ = 10	6 + 2 = ___
___ + 1 = 10	___ + 0 = 6	10 - ___ = 6	___ - 2 = 7
8 - 6 = ___	___ - 4 = 3	9 - ___ = 6	10 - ___ = 9
___ - 5 = 3	___ - 0 = 10	9 - 3 = ___	7 - ___ = 2

48

Recording Sheet

What's Missing?

Player 1: _____		Player 2: _____	
10		10	
9		9	
8		8	
7		7	
6		6	
5		5	
4		4	
3		3	
2		2	
1		1	

Teacher Page

Odds and Evens

Help children recognize if a number is odd or even with this quick and easy activity. Children roll a die and color in that number of circles. If every circle has a partner, then the number is even. Otherwise, the number is odd. Have children work on this activity independently or with a partner.

HERE'S HOW

Make double-sided copies of the "Odds and Evens" activity page. Give each child an activity page and a die.

Display a copy of the activity page on the board and model how to do the activity. Roll the die and write the number on the activity sheet. Then color in the corresponding number of circles. Emphasize to children that when they color, they should start with the top left circle, then go down to the circle below it, then move to the top of the next column, then down, and so on.

Explain to children that if every circle has a partner, then the number is even. If there is a circle without a partner, then the number is odd. Have children fill in both sides of their activity sheet.

MATERIALS

- Odds and Evens activity page (page 51)
- dice (9-sided, 12-sided, or 20-sided, depending on children's abilities)
- crayons
- classroom projection system

EXIT TICKET

Tell each child either "odd" or "even." Have the child give you a number that fits the criteria.

Alternatively, you can say a number from 1 to 120 and have the child tell you if the number is odd or even.

Activity Page

Name: _____ Date: _____

Odds and Evens

Roll the die. Write the number below. Color that number of circles. Is the number odd or even?

Number Rolled	Counters	Circle: Odd or Even
		Odd
		Even
		Odd
		Even
		Odd
		Even
		Odd
		Even

Play & Learn Math: Addition & Subtraction © Mary Rosenberg, Scholastic Inc.

Teacher Page

Add and Subtract Odds and Evens

With this game, help children recognize patterns when adding or subtracting combinations of odd and even numbers.

HERE'S HOW

Divide the class into small groups or pair up children. Provide each group with a pair of dice, then give each child a copy of the "Add and Subtract Odds and Evens" recording sheet and a pencil.

Display the recording sheet on the board and model how to play the game. Roll both dice. Using the numbers rolled, make either an addition sentence or a subtraction sentence on the recording sheet. Write the math sentence in the appropriate column. For example, say you roll a 3 (odd) and a 4 (even). You can write either 3 + 4 = 7 in the "Odd + Even" column or 4 − 3 = 1 in the "Even − Odd" column.

Children take turns rolling the dice and filling their recording sheet. The first child to complete a column wins the round. After playing the game, have children answer the question at the bottom of the page.

MATERIALS

- Add and Subtract Odds and Evens recording sheet (page 54)
- 2 dice (6-, 9-, or 12-sided) for each pair or small group of children
- pencils
- classroom projection system

EXIT TICKET

Tell each child "two odds," "two evens," or "one odd and one even." Have the child share an addition or subtraction problem that fits the criteria you gave.

Game Directions

Add and Subtract Odds and Evens

Add and subtract odd and even numbers. Do you see any patterns?

What to Do

1 Roll both dice.

2 Look at the numbers. Make an addition or subtraction sentence using the numbers.

3 On your recording sheet, look for the strategy you used. Write the math sentence under that strategy.

4 The next player takes a turn.

5 Continue taking turns. The first player to complete a column on his or her recording sheet wins.

Players: 2 to 4

You'll Need

★ Add and Subtract Odds and Evens recording sheet (for each child)

★ 2 dice

★ pencil (for each child)

Recording Sheet

Name: _____ Date: _____

Add and Subtract Odds and Evens

Odd + Odd	Even + Even	Odd + Even	Odd − Odd	Even − Even	Odd − Even	Even − Odd

What do you notice about adding and subtracting odd and even numbers?

_____

_____

Teacher Page

Make It Home

Children practice adding and subtracting bigger numbers in this fun two-player game. Using a 120 Chart and a die, players race on the chart to get to their own respective finish lines (120 for addition and 1 for subtraction).

HERE'S HOW

Pair up children. Provide each pair with a copy of the 120 Chart and the "Make It Home" recording sheet, plus two different-color counters and a die. Have partners decide who will do addition facts and who will do subtraction facts. (Players will switch roles.)

Model how to play the game. On the 120 Chart, the Addition player places his counter on 61, while the Subtraction player places her counter on 60. The Addition player rolls the die and adds that number to 61. He writes the number and sum in the recording sheet, then moves the counter to that sum. Next, the Subtraction player rolls the die and subtracts that number from 60. She also writes the number and difference in the recording sheet, then moves the counter to that difference.

Players continue taking turns until either the Addition player reaches 120 or the Subtraction player reaches 1. The first player to reach his or her goal wins.

Have partners switch roles and play again.

MATERIALS

- 120 Chart (page 57)
- Make It Home recording sheet (page 58)
- dice (6-sided for beginner children, 9- to 12-sided for more advanced children)
- different-color counters
- pencils

EXIT TICKET

Give each child a blank sheet of paper and a starting number on the 120 Chart. (For example, all of the children might start at 83.) Roll the die and have the child either add or subtract the number from the starting number. Have children write the math sentence on their paper to be used as their Exit Ticket.

Game Directions

Make It Home

Add or subtract numbers on the 120 Chart. Who will reach home first?

What to Do

1. Decide who will do addition and who will do subtraction.

   - The Addition player places his counter on the number 61 on the 120 Chart.

   - The Subtraction player places her counter on the number 60.

2. The Addition player rolls the die.

3. Write the number rolled on the recording sheet. Then add it to 61.

4. The Addition player moves his counter to the new sum.

5. The Subtraction player takes a turn rolling the die.

6. Write the number rolled on the recording sheet. Then subtract it from 60.

7. The Subtraction player moves her counter to the new difference.

8. Players continue taking turns.

9. If the Addition player reaches 120 first, he wins. If the Subtraction player reaches 1 first, she wins.

10. Players switch roles and play again.

Players: 2

You'll Need
- ★ 120 Chart
- ★ Make It Home recording sheet
- ★ die
- ★ counter (different color for each child)
- ★ pencil (for each child)

Number Chart

120 Chart

1	2	3	4	5	6	7	8	9	10
11	12	13	14	15	16	17	18	19	20
21	22	23	24	25	26	27	28	29	30
31	32	33	34	35	36	37	38	39	40
41	42	43	44	45	46	47	48	49	50
51	52	53	54	55	56	57	58	59	(60)
(61)	62	63	64	65	66	67	68	69	70
71	72	73	74	75	76	77	78	79	80
81	82	83	84	85	86	87	88	89	90
91	92	93	94	95	96	97	98	99	100
101	102	103	104	105	106	107	108	109	110
111	112	113	114	115	116	117	118	119	120

Recording Sheet

Make It Home

Addition Facts	Subtraction Facts
61	60
+	−
+	−
+	−
+	−
+	−
+	−
+	−
+	−
+	−

Teacher Page

A Flip of the Cards

Introduce children to arrays and repeated addition with this simple card activity. You may want to have children work with a partner at first. Then when they've had more experience, children can work on their own.

HERE'S HOW

Pair up children. Provide each pair with three or four sets of number cards from 1 to 5. Give each child a copy of the "A Flip of the Cards" recording sheet.

Display a copy of the recording sheet on the board and model how to do the activity. Shuffle the cards and stack them facedown on the table. Turn over the top two cards and draw the matching array on the recording sheet. Then write the two repeated addition sentences that describe the array—both vertically and horizontally.

For example:

4 + 4 + 4 = 12

3 + 3 + 3 + 3 = 12

If children are working in pairs, have them take turns picking cards, drawing the array, and writing addition sentences.

MATERIALS

- A Flip of the Cards recording sheet (page 61)
- 3 or 4 sets of number cards from 1 to 5* (for each pair of children)
- classroom projection system

* You can make these with index cards or use playing cards.

EXIT TICKET

Share a word problem with each child. For example: *Dad has 3 cookie trays. He put 5 cookies on each tray. How many cookies in all?* Have the child tell you the addition sentence: 5 + 5 + 5 = 15. More advanced children can share the multiplication sentence: 3 × 5 = 15.

Independent Practice

A Flip of the Cards

Pick two numbers. Then draw the array and find the sum.

What to Do

1. Shuffle the cards. Stack them facedown on the table.

2. Turn over the top two cards.

3. Write the numbers on the recording sheet.

4. Draw the matching array.

5. Describe the array. Write two repeated addition sentences. For example:

You'll Need
- ★ A Flip of the Cards recording sheet
- ★ 3 or 4 sets of number cards 1 to 5
- ★ pencil

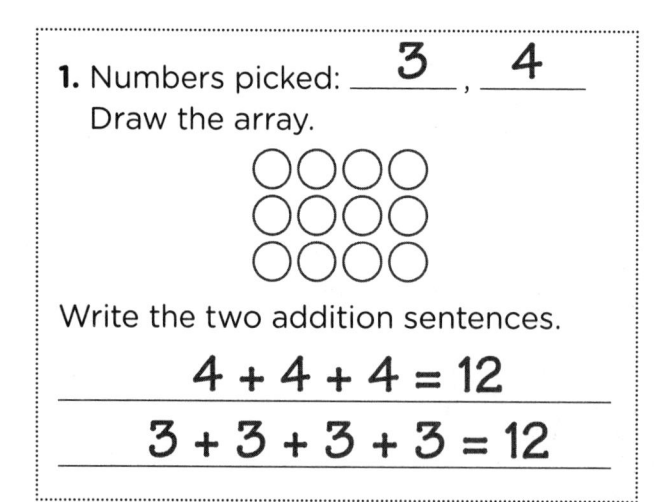

---

1. Numbers picked: __3__ , __4__
   Draw the array.

   ○○○○
   ○○○○
   ○○○○

   Write the two addition sentences.

   4 + 4 + 4 = 12
   3 + 3 + 3 + 3 = 12

---

6. Repeat Steps 2 to 5 a few more times.

Recording Sheet

Name: _____ Date: _____

A Flip of the Cards

1. Numbers picked: _____ , _____
   Draw the array.

   Write the two addition sentences.
   _____
   _____

2. Numbers picked: _____ , _____
   Draw the array.

   Write the two addition sentences.
   _____
   _____

3. Numbers picked: _____ , _____
   Draw the array.

   Write the two addition sentences.
   _____
   _____

4. Numbers picked: _____ , _____
   Draw the array.

   Write the two addition sentences.
   _____
   _____

5. Numbers picked: _____ , _____
   Draw the array.

   Write the two addition sentences.
   _____
   _____

6. Numbers picked: _____ , _____
   Draw the array.

   Write the two addition sentences.
   _____
   _____

Teacher Page

Duck Gallery

To play this game successfully, children must be able to fluently add and subtract within 20 using mental strategies. A fun skill-building game for math centers or independent practice.

HERE'S HOW

Pair up children and provide each pair with a 6-, 9-, or 12-sided die (depending on their abilities). Give each child a copy of the "Duck Gallery" recording sheet and a pencil.

Display a copy of the recording sheet and model how to play the game. Roll the die. Then use the number in an addition or subtraction problem in which the answer is one of the numbers on the ducks. Write the addition or subtraction sentence under the corresponding duck. For example, say you roll an 8. Sample sentences you can write include:

8 + 12 = 20 (Write this under the "20" duck.)

8 − 4 = 4 (Write this under the "4" duck.)

Players take turns rolling the die and creating a math sentence. Each duck can be used only once. The first player to use all 20 ducks wins the game.

MATERIALS

- Duck Gallery recording sheet (page 64)
- 6-, 9-, or 12-sided dice
- pencils
- classroom projection system

EXIT TICKET

Write the numbers 1 to 20 on index cards and shuffle the cards. Turn over two cards and have children tell the addition sentence (if the sum is less than 20) or the subtraction sentence.

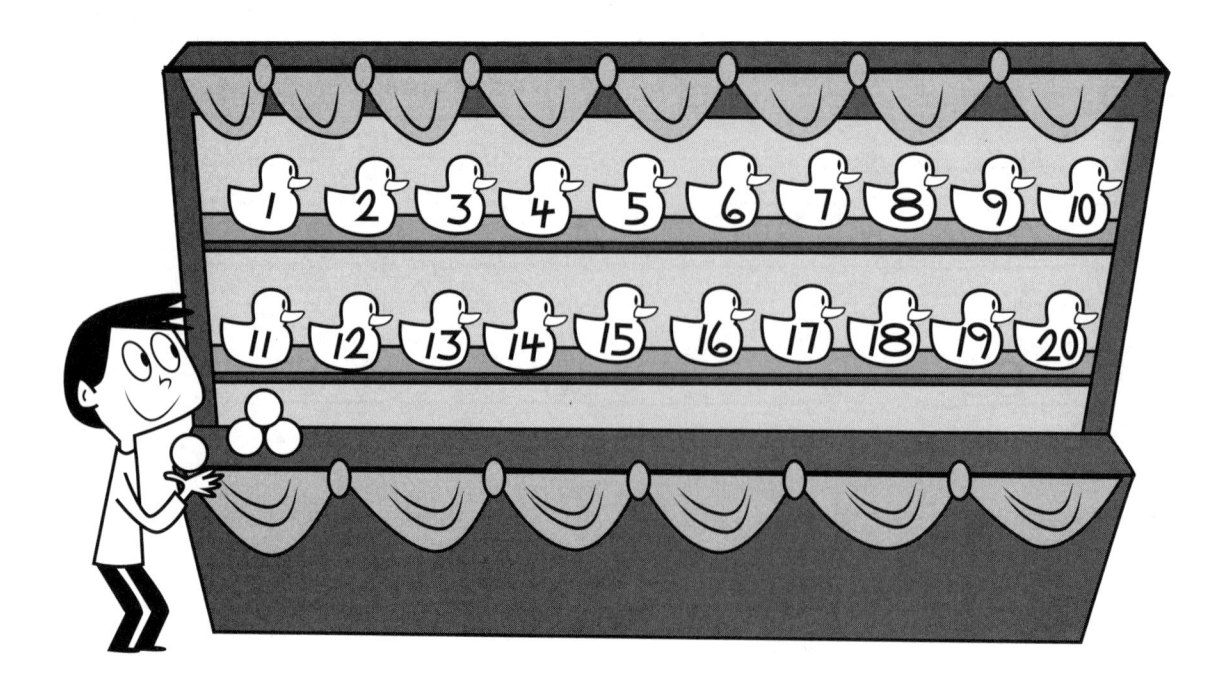

Game Directions

Duck Gallery

Make a math sentence. The answer should match a number on a duck. The first player to get all 20 ducks wins.

What to Do

1. Roll the die.

2. Use the number in an addition or subtraction sentence. The answer should match a number on a duck.

3. Write the math sentence under the duck on your recording sheet. You can use each number duck only once.

4. The next player takes a turn.

5. Continue taking turns. The first player to use all 20 ducks wins.

Players: 2

You'll Need
- ★ Duck Gallery recording sheet (for each child)
- ★ die
- ★ pencil (for each child)

Recording Sheet

Name: _____ Date: _____

Duck Gallery

1	2	3	4	5
6	7	8	9	10
11	12	13	14	15
16	17	18	19	20